Design Knowledge

Design Knowledge

A Visual Guide

Rojin S. Vishkaie

To order additional copies of this book, contact:
Xlibris
1-888-795-4274
www.Xlibris.com
Orders@Xlibris.com
697496

CONTENTS

For my parents

Audience

This book is primarily aimed at undergraduate and graduate students, as well as academic researchers and practitioners. This book is naturally intended for industrial designers, graphic designers, media and communication designers, urban designers, fashion designers, architects, sociologists, computers scientists, and engineers. This book is a useful resource for the design research process of products, artifacts, services, systems, media, interfaces, buildings, and fashions.

What Is Design Knowledge?

The author of *The Reflective Practitioner: How Professionals Think in Action*, Donald Schön (1930–1997), the influential thinker in developing the theory and practice of reflective professional learning, proposed to search for "an epistemology of practice implicit in the artistic, intuitive processes which some practitioners do bring to situations of uncertainty, instability, uniqueness, and value conflict." Despite the doctrine of the design science movement proposed by his positivist predecessors in twentieth century, Schön offered a constructivist paradigm in which interdisciplinary design studies involve creative activity of making the artificial artifacts. Design as an independent discipline, which owns rigorous intellectual culture, can form knowledge through reflecting on the creation of those artifacts. Thus this book tries to reemerge the intellectual dimensions of the design knowledge.

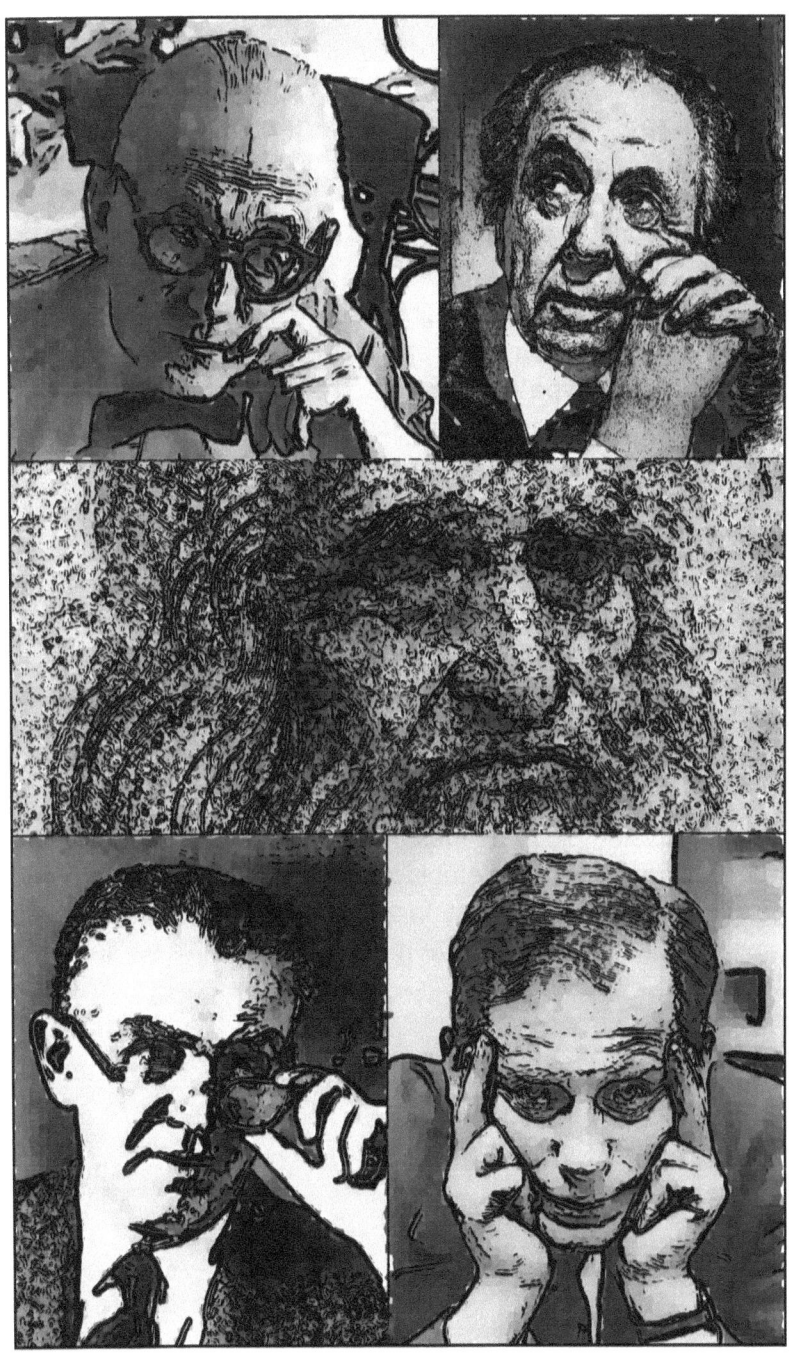

Epistemology

Qualitative and Quantitative Research show that research is divided into two research parts. The distinction between the qualitative research and quantitative research is related to the appropriate methods for each type of research. The level of distinction occurs between subjectivist, or constructionist, research and objectivist, or subjectivist, research.

Subjectivist, or constructionist, research is associated with qualitative methods, which ascertain meanings in people's lives and daily contexts. However, objectivist, or positivist, research is associated with quantitative methods, which deal with absolute truth and valid and generalizable conclusions. Furthermore, qualitative research generates suggestive outcomes; however, quantitative research generates conclusive outcomes.

Moreover, epistemology defines the philosophical grounding of the research. Thus, to be epistemologically consistent, research needs to be subjectivist or objectivist. In addition, subjective meanings can be approximated from objective meanings, and vice versa.

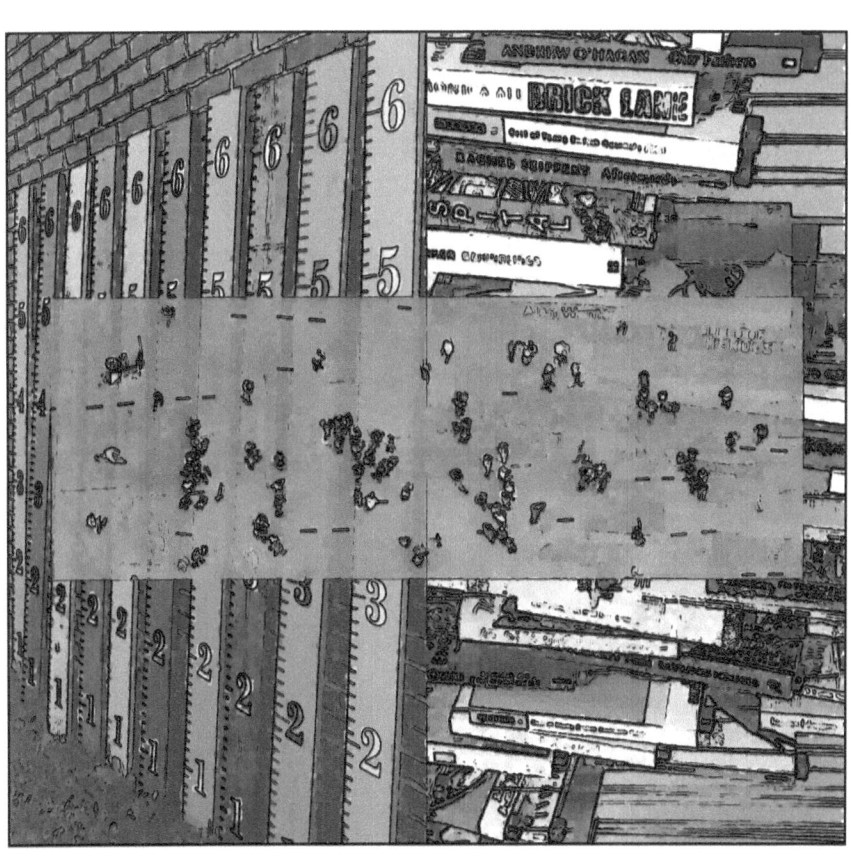

Theoretical Perspective

Postmodernism approach in qualitative research determines an open-ended and evolving dialogue based on imaginative, interpretative, and descriptive practical reasoning, which is removed from theory-centered reasoning. Positivism relies on reliability and validity criteria as an enabling condition in qualitative research and argues that "no criterion can ever be independent of our own construction of it." However, postmodernism moves beyond criteriology in which skepticism is advocated instead of certainty, also understanding instead of knowledge. Thus, the notion of "probable truth" in a postmodernist perspective recognizes that "no set of standards can ensure confidence that research findings are indeed entirely valid."

Methodology

Ethnography is descriptive and interpretive study of beliefs, patterns of values, meanings of behaviors and languages, and the interaction among members of the culture-sharing group. Ethnography involves observations and interviews with the group of participants. Some forms of ethnography include life history, auto-ethnography, feminist ethnography, and visual ethnography. However, two popular forms of ethnography are critical ethnography and realist ethnography. The critical ethnography studies the issues of power, empowerment, inequality, inequity, and victimization. The realist ethnography is an objective perspective on the situation, which reports objectively on facts.

Cognitive Ethnography is the attempt to study the pattern of shared understanding and its evolution over time in relation to individual minds. Cognitive ethnography focuses on events and seeks to understand what individuals know, how they acquire the knowledge, and how they use the knowledge to do what they do. It is also necessary to understand situated human cognition, how individual minds process information, and how the information to be processed is coordinated with social and material actions. Cognitive activities consist of internal and external resources. Understanding the context of actions in relation to internal and external resources provides meanings to social and material actions.

Distributed Cognition, as the most socially oriented activity, focuses on understanding the whole environment where users and artifacts, tools, and technologies interact within the course of working. The characteristics of the work process create a unique cognitive ecology that includes the users and artifacts they use in their functional system. The distributed cognition approach is used to touch upon the cognitive processes that emerge from interaction among internal (mental) and external (cultural) resources within the course of working.

Distributed cognition can provide an understanding of the users' work settings through exploration of the work process. This could lead to a set of concepts that describes the complexities and challenges within the work process. The distributed nature of cognitive phenomenon can discover the context within the work process through forming a partnership with users, and therefore focusing on artifacts, and internal and external representations of such a process. These issues can further be described in terms of the work's communication and technical pathways, such as poor communication, ineffective propagation of information, and coordination of different technologies. This results in a contextual description of the users' work that stresses information and its propagation through the cognitive system of the work process.

However, integrating various contextual perspectives and influences of the users and the artifacts, tools, and technologies into the participatory nature of the work process is challenging. Distributed cognition can elaborate on functional relationships between elements that participate in the work process where a group of minds without a unified mind-set seek to store their distributed knowledge in external artifacts, tools, and technologies. The work process is a cognitive functional system, in which the boundaries of the unit of analysis for cognition are defined by the functional group rather than considering individual minds. This functional group consists of the users and their organized processes and interactions with one another and with artifacts. The relation between these distributed cognitions needs to be contextualized and coordinated effectively to accomplish a new functional skill in a dynamic functional system, which could lead to a potential outcome within the work process.

Research-in-the-Wild aims at studying work-in-context along with concrete details of cooperative work. In contrast to abstract, recognizable and in general theoretical detailed descriptions of workaday activities, these methods offer a radical approach to the study of work in situ. Such subtleties of socially constructed work practices provide an opportunity to understand what is really going on in the course of the work, what the problem really is, and what the possible design solutions might be.

Case Study is a type of design in qualitative research, which focuses on one or more cases within a bounded system to explore and understand the context of an issue or problem. Case study research uses multiple sources of information to collect in-depth and detailed data. These sources include interviews, observations, audio-visual material, documents, and reports. The qualitative case studies are distinguished by the size of the bounded case, such as an individual, individuals, or a group. In addition, the qualitative case studies may also be distinguished in terms of their intents, such as intrinsic case study, instrumental case study, and collective case study. Intrinsic case study focuses on unusual or unique cases. Instrumental case study focuses on an issues or concern, and then the bounded case is selected. Collective case study focuses on multiple case studies.

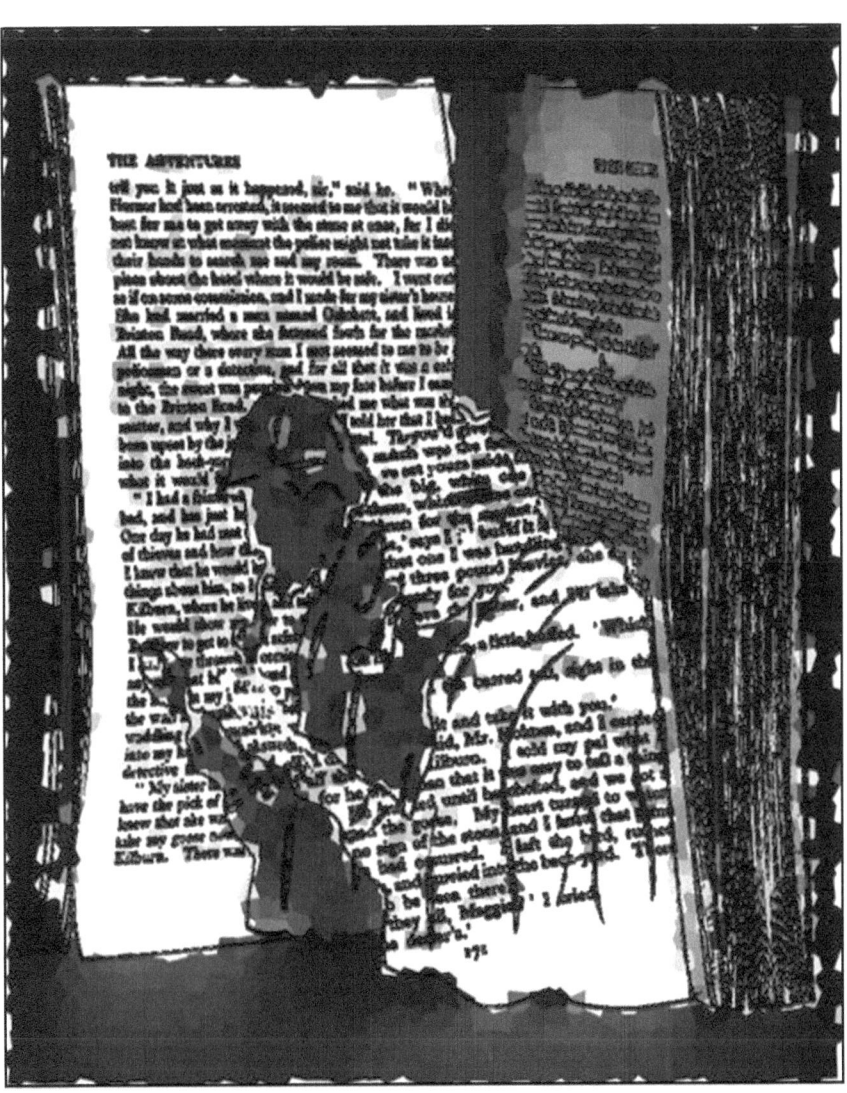

Contextual Design (CD) originates from the organizational challenges of designing general purpose systems. Contextual design considers principles in psychology, anthropology, sociology, and hermeneutics in designing computer systems. Contextual design aims at identifying an appropriate way of gathering helpful information about participants involved in similar work practices and organizational challenges. Contextual design tends to transform individuals' and teams' work practices through the use of computer systems, including hardware, software, services, and support. To achieve this, contextual design provides an understanding about the user's work practices. Contextual design was developed by Hugh Beyer and Karen Holtzblatt in 1998.

Rapid Contextual Design (RCD) is a practical guide for practitioners of the conceptual design technique. Depending on the type of rapid contextual design project, some or all the steps in rapid contextual design can be used. These steps are referenced in a deeper discussion of the contextual design technique. These steps include contextual inquiry, interpretation sessions and work modeling, model consolidation and affinity building, personas, visioning, storyboarding, user environment design (UED), and paper prototyping and mock-up interviews. Furthermore, the main goal of rapid contextual design is to reduce the time that is needed to include the user data in the existing design processes within companies. Moreover, rapid contextual design focuses on recognizing which steps are unnecessary and could be skipped.

Participatory Design is an approach to design that attempts to include users in various stages of the codesign process to create a useful design output that addresses the users' needs. Thus, users cooperate with designers, researchers, and stakeholders within various stages of the design process, including exploration, creation, and evaluation. Furthermore, participatory design has a political dimension of user empowerment and democratization. Moreover, distributed participatory design is a design approach that supports participation of distributed users, designers, researchers, and stakeholders within the cooperative design process. Participatory design was used in several Scandinavian countries during the 1960s and 1970s, and it was developed based on action research.

User Experience Design (UXD) is a process that attempts to enhance users' satisfaction through improving ease of use, simplicity, and desirability of products, services, and environments. Traditionally, user experience design was used in the fields of human factors and ergonomics; however, this method has a recent connection with human-computer interaction and user decentered design. User experience design focuses on humanizing products, systems, and environments through understanding the users' unspoken needs; the tools, artifacts, and technologies they use; and the interaction among them.

User-Centered Design (UCD) is a multistage process that focuses on recognizing users' needs, expectations, and limitations within the design processes. User-centered design is a process of iterative exploration, creation, evaluation, and post evaluation of a product, system, or environment. This method could help to gain insights about at various phases of the design life cycle. User-centered design focuses on the rhetorical situation that include audience, purpose, and context. In addition, the main analysis tools in user-centered design are persona, scenario, and use case.

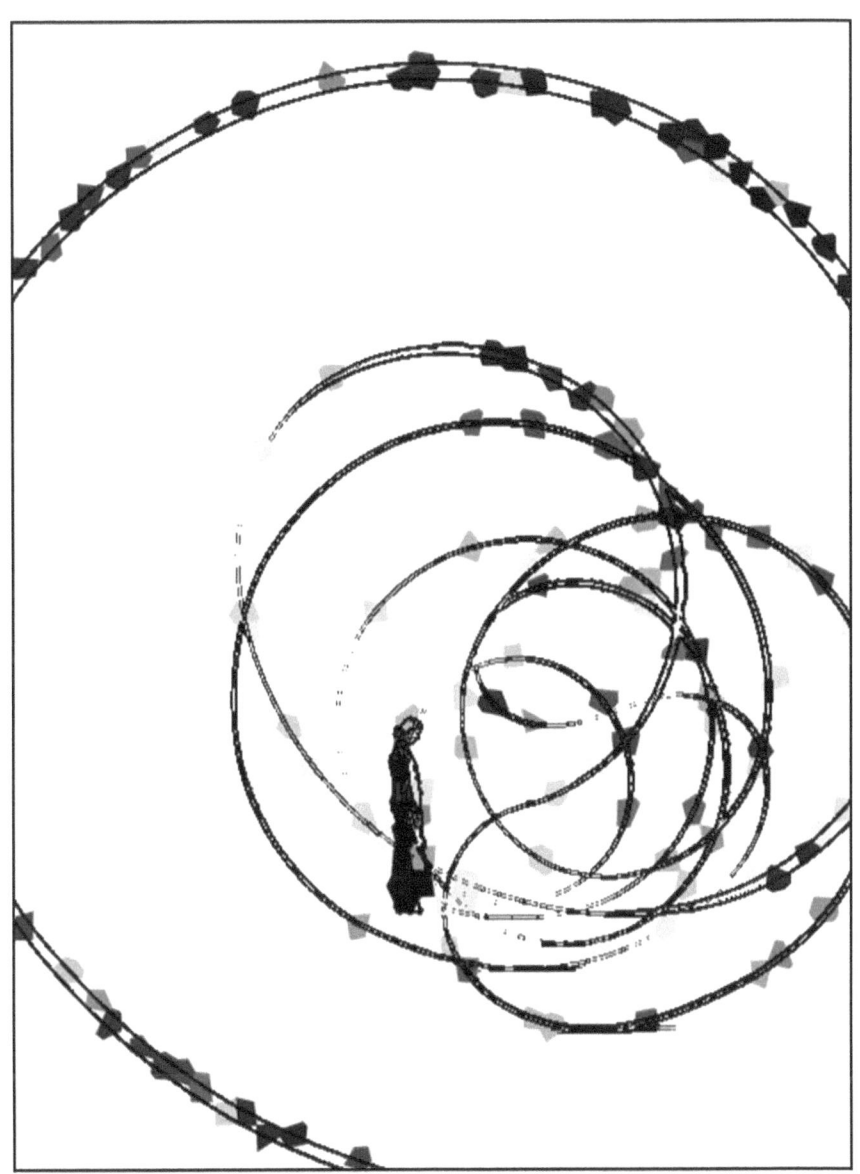

Experience Sampling Method (ESM) is a research methodology that inquires participants to provide samples of their ongoing behavior in the place and time of occurrences. In this method, participants' reports are contingent upon a signal, preestablished interval, or occurrence of an event. Signal contingent involves signaling participants via a phone call, beeper, and stopwatch within a fixed period of time. When participants are signaled, they record their feelings, thoughts, activities, and locations. Interval contingent assigns preestablished intervals for participants to report events. For example, participants fill out a survey of their daily activities before finishing a workday in the evening. Event contingent determines a key event based on a research project's objectives. When the event occurs, participants record the detail of the event. Participants can be given a journal including psychometric scales or open-ended questions to record temporal things. Experience sampling method was developed by Reed Larson and Mihaly Csikszentmihalyi.

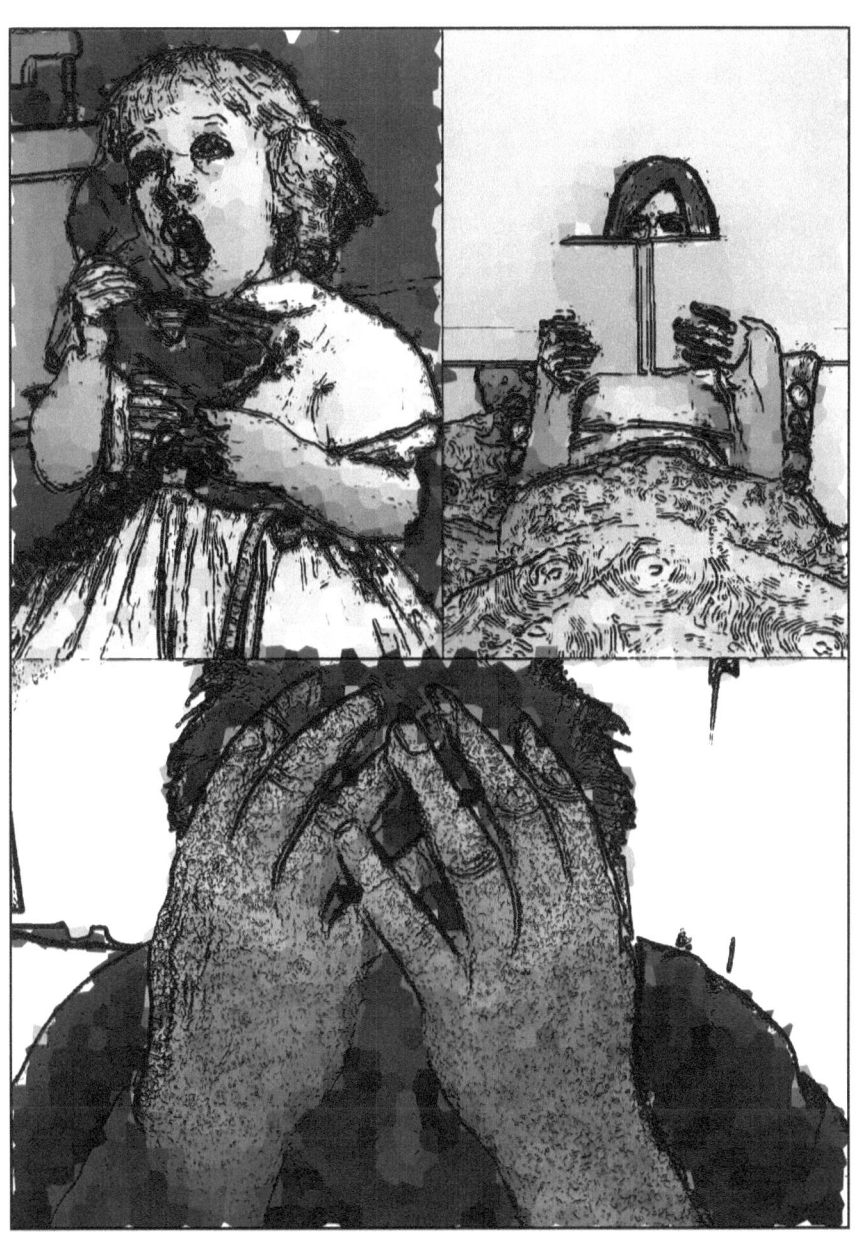

Empathic Design is a user-centered design process that focuses on users' feeling toward products, systems, and environments. Empathetic design attempts to identify the untapped users' needs and expectations to create new products, artifacts, and technologies that offer suggestions to users' difficulties and inefficiencies where users are unfamiliar with the new possibilities because of lack of awareness about new technological potentials. As opposed to traditional market research, which involves surveys and questionnaires, empathic design technique focuses on users' observation as an effective method of data gathering. Ideally, the observation session should include designers, engineers, ethnographers, and anthropologists to provide a better insight about the unarticulated users' needs. Observation sessions should then follow reflection and analysis, brainstorming, and prototyping. Empathic design technique was first adopted by automotive and electronic product manufacturing industry.

Universal Design (also known as inclusive design) focuses on designing useful and aesthetically pleasing products, services, and built environments that are barrier-free and accessible to everyone including people with disabilities, people without disabilities, and older people. There is a growing interest in universal design as the life expectancy increases and modern medicine advances. Universal design has a root in barrier-free design and accessible design. Some of the principles of universal design include simple and intuitive, perceptible information, flexibility in use, equitability in use, tolerance for error, and low physical effort. In addition, Design for All (DfA) presents the design philosophy that ensures products, services, and built environments are easy to use, accessible, and affordable for everyone regardless of their age, ability, or status in life.

Sustainable Design, also known as environmental design, is the process of creation of products, services, and built environments that comply with the socioeconomic and ecologically sustainable philosophical perspective. The goal of sustainable design is to eliminate negative environmental impact by increasing products, services, and built environments longevity and change of human behavior toward natural resources. Some of the sustainable design principles include low-impact materials, energy efficiency, emotionally durable design, design for reuse and recycling, biomimicry, renewability, and robust eco-design. Sustainable design philosophy can be applied in the fields of architecture, landscape architecture, urban design, urban planning, engineering, graphic design, industrial design, interior design, fashion design, and human-computer interaction.

Data Collection Methods

Interview is one of the most popular forms of qualitative, ethnographic form of data collection, ranging from close-ended to open-ended. There are three types interviews, such as telephone interviews, focus group interviews, and one-on-one interviews. Purposeful sampling should be used to identify interviewees. It is also recommended to record the interviews. The central interview questions should be predefined to answer the research questions. In addition, pilot study is a good method to refine the interview questions and procedure. Consent from the interviewee should be obtained to participate in the study.

Contextual Inquiry is part of the methodology of contextual design that incorporates a user-centered design and ethnographic methods for open-ended interviews with users. Three principles of contextual inquiry process are context, partnership, and focus.

Context provides an understanding of the users' work in the course of working through observation of the users during their ongoing work and through ongoing dialogue with them while they use work artifacts in their actual work environment.

Partnership with the users through dialogue needs to be established to direct the area of concerns, along with empowering users to lead the conversation and articulate the nature of their work and subtleties pertaining to their activities within the course of working.

Hence, the focus of the inquiry is to uncover users' experience of work and their use of artifacts, tools, and technologies within the course of working and to develop a series of work models including flow, sequence, artifact, cultural, and physical models of the work.

Survey is a method for collecting data about participants' thoughts, opinions, interests, and feelings. A survey consists of a set of predetermined and closed questions, which are given to representative sample users, generalizing the results and describe the larger population. Web-based surveys, mail surveys, telephone surveys, and one-on-one surveys are four main methods of conducting users' surveys.

Likert scale and semantic differential scale are bipolar, psychometric scaling methods that are commonly used in research that employs questionnaire. The typical, discrete Likert scale has seven items, including (1) strongly agree, (2) agree, (3) somewhat agree, (4) no opinion, (5) somewhat disagree, (6) disagree, and (7) strongly disagree. In addition, the typical semantic differential scale has seven items, or seven bipolar pairs of adjectives. For example, there are seven separate positions on a scale that are numbered 1 through 7, with 1 being "unimportant" and 7 being "important." Furthermore, consent from the participants should be obtained to take part in the study.

Observation is one of the most popular forms of qualitative, ethnographic form of data collection, ranging from nonparticipant to participant. Nonparticipant and participant observations provide an approach of changing researcher's role from an outsider to insider through the course of observation sessions. Who, what, when, and how long to observe need to be identified. Recording method, including audio-visual recoding or note taking, should also be identified. Moreover, the objectives of the observations should be determined. Furthermore, consent from the participants should be obtained to take part in the study.

Focus Group is qualitative, ethnographic form of data collection where interviewees are asked to speak about their opinions, perceptions, attitudes, and beliefs toward products, services, or built environments. Focus group is an advantageous method when the interaction among the interviewees provides the better information, when there is a limited time to collect information, and when interviewee is hesitant to speak or share opinions within one-on-one interviews. Furthermore, consent from the interviewees should be obtained to participate in the study.

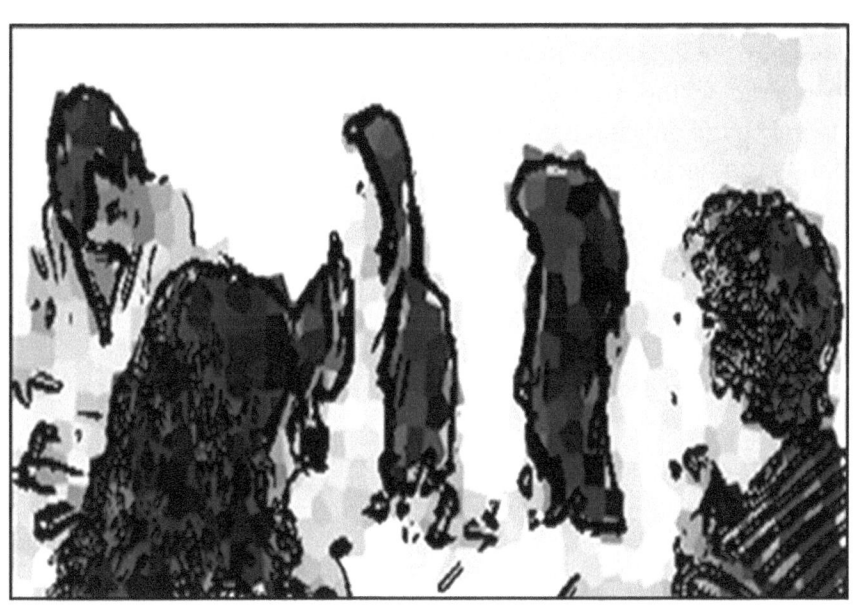

Pilot Study is conducted as a small-scale exploration designed to gather information prior to a larger study and to evaluate the feasibility of the further research. Pilot study can help clarifying the aspects that needs to be measured in the process of experimental research. In addition, the results of the pilot study can be useful in improving the quality and efficiency of the further research by revealing the deficiencies in the study design. Also useful information can be gathered for the later iterations of the research.

Thinking Aloud is a useful method for evaluation of a system that attempts to invite users to continuously think out loud and verbalize their thoughts about the system. The thinking aloud method provides an understanding of users' conceptions and misconceptions about the system. The advantage of using the thinking aloud method is to show what the users are doing while they making comments regarding what they like and dislike about the system.

However, the disadvantage of using the thinking aloud method is that some users have difficulty verbalizing their thoughts as this method seems unnatural to them. In addition, thinking aloud method may provide an erroneous understanding of the system evaluation as a result of accepting users' comments uncritically. There are three types of thinking aloud methods; these include the following: Constructive interaction involves two users to evaluate the system. Retrospective testing, which allows collecting additional information about the users' interaction with the system, is an audio-video recording that has been made of the session. And coaching method involves the experimenter to navigate the user in the right direction while using the system. Traditionally, thinking aloud method has been used as a psychological research method; however, it has increasingly been used for the evaluation in the field of design and human-computer interaction.

Five Ws (Wh-words) are questions that are used for basic date gathering in research, journalism, and police investigations. These questions tend to provide a complete story about the context under study since they cannot be answered with yes or no. These questions include the following: Who takes action? What action are they taking? Where are they taking action? When are they taking action? Why are they taking action?

Triangulation uses two or more methods to ensure that all aspects of a phenomenon under study have been investigated from multiple perspectives. To enhance the quality of the research, triangulation can be used to provide various methods of data collection, such as interviews, observations, surveys, focus groups, and walkthroughs of prototypes. To complete the understanding of the concepts, a range of data sources, such as a variety of individuals, spaces, and times, needs to be chosen for interviews, observations, surveys, focus groups, and walkthroughs of prototypes. This can ensure that final representation of the data confidently and accurately reflected the experience.

Data Recording Methods

Audio-Video Materials are preferred approaches to data collection in qualitative research and can be used in interviews, observations, and focus groups. Audio-video materials include photographs, audio and video materials, and digital archives, such as photo elicitation. Audio-video materials, in conjunction with an interview or observational protocol, which is a predesigned technique (e.g., a form), can be used to record data collected.

Written Materials and Physical Artifacts are also preferred approaches to data collection in qualitative research and can be used in interviews, observations, and focus groups. Written materials and physical artifacts include researcher field notes, letters, journaling (memoing), diaries, autobiographies, biographies, stories of families, documents, maps, sketches, and personal-social artifacts. Written materials and physical artifacts, in conjunction with an interview or observational protocol, which is a predesigned technique (e.g., a form), can be used to record data collected.

Interpretation Methods

Affinity Diagramming is an analytic method for analyzing the data through structuring ideas from a set of unstructured ideas. The affinity diagramming derives from seven quality processes, also known as K-J methods. This is a manual, noncomputational technique commonly used in contextual design for illustrating the participants' problem context by focusing more on the practical side of the design process. It aims at interpreting and organizing the qualitative data collected during interviews, observations, and surveys through inductive, bottom-up reasoning. It is also used for analyzing the data by dividing the data into elements. Then each element is categorized in a meaningful way with an appropriate granularity. Categories of the data that emerged from participants' insights and ideas are defined by the goals of the study. This also helps organize the data into a hierarchy presenting common patterns and structures. These patterns and structures emerge from the data, rather than from a predefined scheme. However, affinity diagramming is a time-consuming, complex, and intensive method, which may require modification in steps based on the requirements of the research.

Cognitive Walkthrough is a method for usability inspection. This method is inspired by the observation that users prefer learning new software through exploration. Cognitive walkthrough is the process of reviewing the proposed design where the designer presents the design to a group of peers for evaluation. In cognitive walkthrough, specific user tasks and sequence of actions that are needed to accomplish a task are evaluated. Furthermore, cognitive walkthrough has the basic rationales as design walkthroughs. Thus, cognitive walkthroughs can help designers to evaluate products, tools, and technologies during the early stages of the design processes.

Heuristic Evaluation is a method that includes a set of principles for usability evaluation of user interfaces. Heuristic evaluation attempts to explain users' likes and dislikes about the design of user interfaces based on the heuristic principles and examiners, thus identifying the usability problems. The heuristics include visibility of the system status; match between system and the real world; user control and freedom; consistency and standards; error prevention; recognition rather than recall; flexibility and efficiency of use; aesthetics and minimalist design; helping users recognize, diagnose, and recover from errors; and documentation. Heuristic evaluation was developed by Jakob Nielson in 1990.

Formative Analysis involves a series of qualitative formal and informal assessments employed throughout the design process. Formative analysis attempts to gather information to assess the effectiveness of the design process and design outcome. Formative analysis focuses on detailed content of the feedback provided the users in order to modify and improve the design process. Formative analysis combines several methods for obtaining information. Some of these methods include focus groups, interviews, observations, and surveys.

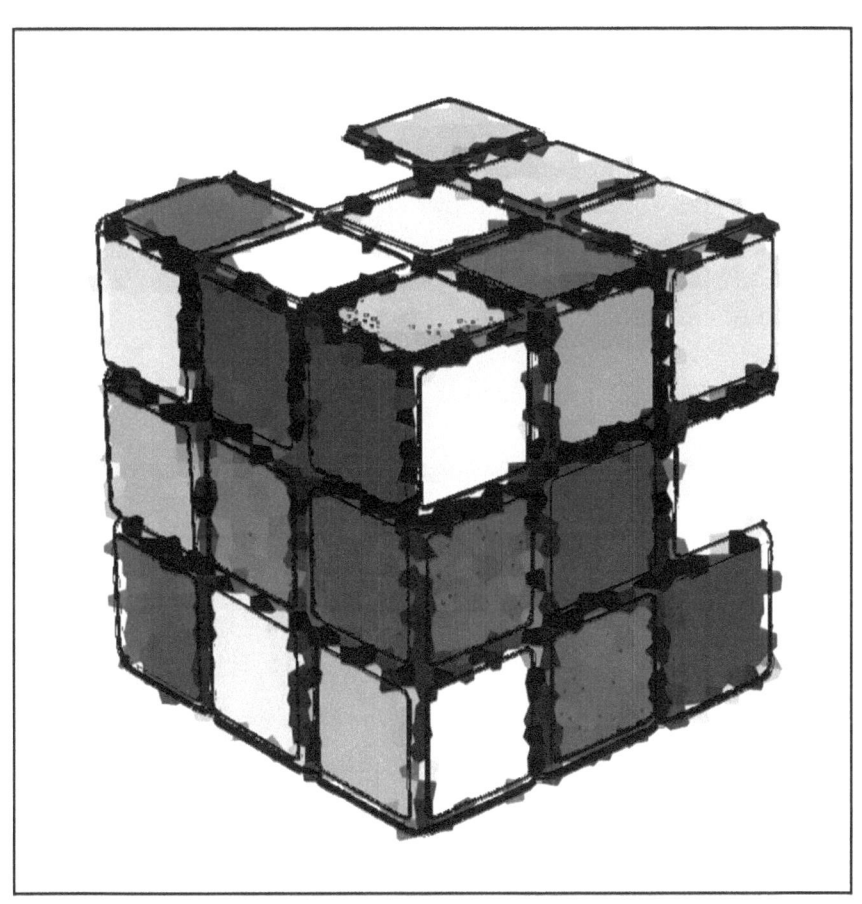

Competitive Analysis can be used to analyze existing products and systems according to usability guidelines. The analysis of existing products and systems provides more realistic analytical results than the analysis of prototypes. In addition, users can perform real tasks on existing products and systems and thus learn more about the expected functionalities and interactions of new products and systems, which are planned to based on the initial analysis of the existing or competing products.

Task Analysis provides initial input to products and systems design. Task analysis studies users' overall goals, intention of performing a task, information they need, and how they approach exceptional circumstances. Task analysis should identify user's model of the task and also identify the weaknesses and strengths of current situation. Furthermore, task analysis can provide a breakdown of the hierarchical structure of the users' tasks and goals. Moreover, task analysis shows users' overall goal, what they want to accomplish, and steps they need to perform to achieve their goals. In addition, as part of task analysis, users can be observed and interviewed to describe the outstanding successes and failures and changes they would like to make to improve the current products and systems.

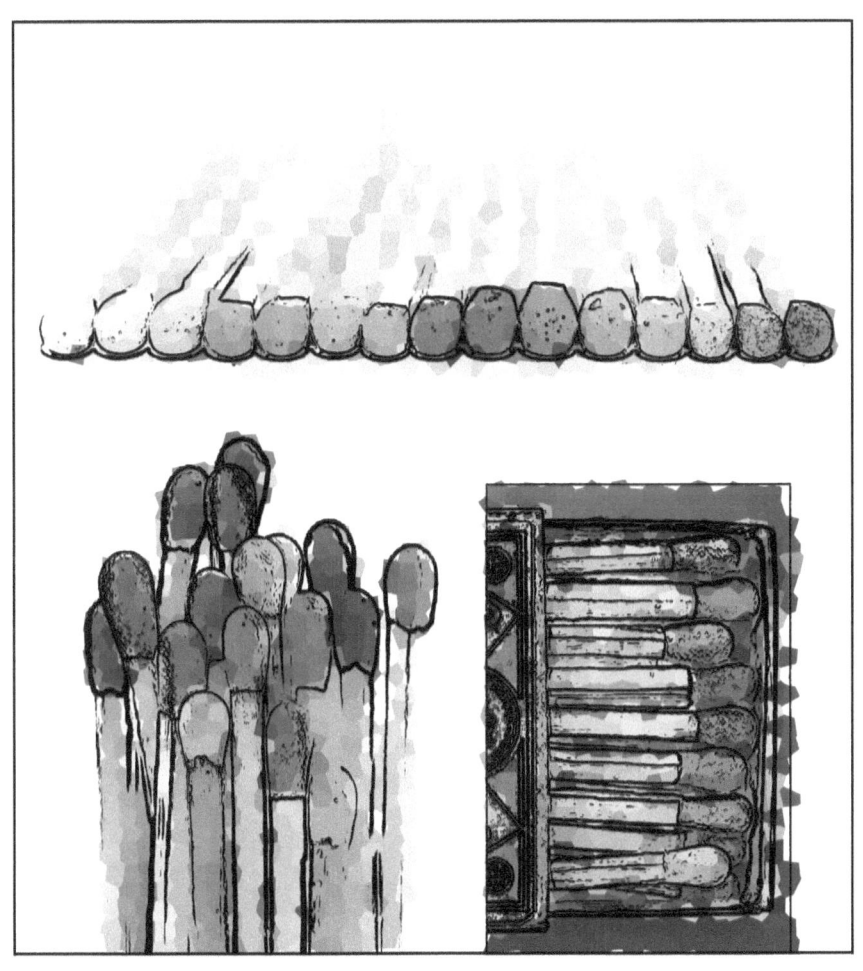

Design Work Models comprise the flow, sequence, artifact, cultural, and physical models, which are used for conceptual grouping of the data. Work models are not intended to represent all the detail but to provide a bird's-eye view of the key aspects of the context under study.

Flow Model uncovers the structure and scope of the work practices, job responsibilities divided across the stakeholders, as well as communication patterns occurring within the work process under study. Job responsibilities are divided up and coordinated according to the needs and intents of the work to ensure that tasks are accomplished. The consolidated flow model included the stakeholders' work practices and responsibilities that are significantly driven by communication around technical activities within the work process.

Sequence Model describes the different steps within a typical work process under study. The sequence model shows the detailed structure of work practices and common strategies undertake by stakeholders across the work process. The sequence model reveals the needs and important aspects needed to accomplish the work and also the orders, strategies, and motivations for doing the work. The consolidated sequence model is developed to understand the intent of activities and also to reveal any breakdowns as well as complexities within the sequence of work practices.

Artifact Model reveals the tools and the technologies used for the work. The consolidated artifact model groups tools and technologies with the same intent, content, structure, and usage that stakeholders use in the course of the work process.

Cultural Model shows common themes and aspects of the culture sharing of stakeholders within the work process under study. The cultural model also demonstrates the interaction of multiple groups of stakeholders while they have different intentions in their actions. The consolidated cultural model reveals the extent of the nature of the work practices.

Physical Model shows the structure of the physical environment surrounding the work under study. The consolidated physical model reveals that the layout of tools and technologies within the physical environment influences the flow of information, communication, coordination, and participation between stakeholders and their workspaces.

Generalizability is generally defined as a set of characteristics that are unrestricted to time and space; hence they must be universal and have pervasive influence. However, qualitative research is characterized by the unique nature of every research situation; thus, findings can be somewhat generalized to certain other disciplines, such as industrial design, media design, urban design, architecture, computer science, and engineering.

Furthermore, naturalistic generalizations "reside-in-mind in their natural habitat." In addition, naturalistic generalizations combine formal knowledge of the context under study with the personal experience of the researcher obtained throughout the empirical and design explorations. Moreover, in qualitative research, the data analysis results from the researcher's interpretations, which may be different from other researchers' interpretations. Therefore, the perspective of the researcher for designing the research studies and in analyzing the data is a personal task that influences the results of the research study. Consequently, the researcher's perspective may be different from how other researchers would potentially design the study or analyze the data.

Design

Minimalism is a user-centered design technique for creating interactions, user interfaces, digital media, and artifacts. The notion of minimalism as a means of reductionism focuses on reducing complexity to define different aspects of simplicity and thus to reach a good design in the new genre of interactions, user interfaces, digital media, and artifacts. Inspired by the idea that form follows function, thus minimalism combines both usefulness and aesthetics to create interactions, user interfaces, digital media, and artifacts. In particular, four notions of minimalism can be employed, including the function, structure, architecture, and composition for the design of interactions, user interfaces, digital media, and artifacts.

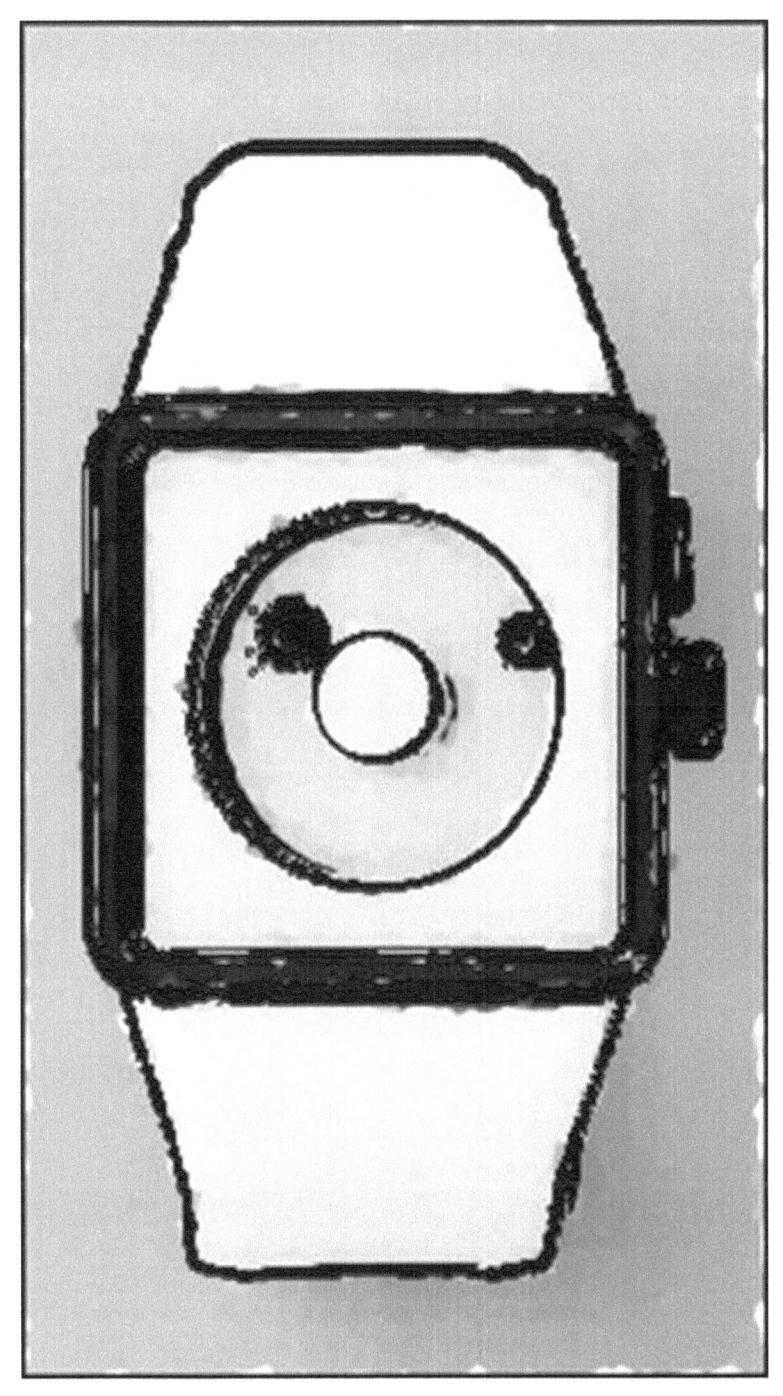

Functional Minimalism intends to reduce complexities for "accessible functionality" in the interactions, user interfaces, digital media, and artifacts. A functionally minimal interaction, user interface, digital medium, and artifact can be simply created to reduce unnecessary functions for the user; thus the interaction, user interface, digital medium, and artifact can be useful only for a single purpose.

Structural Minimalism intends to reduce complexities of the "perceived access structure" of the interactions, user interfaces, digital media, and artifacts. A structurally minimal interaction, user interface, digital medium, and artifact can be created in a manner that does not involve the user with unnecessary navigation.

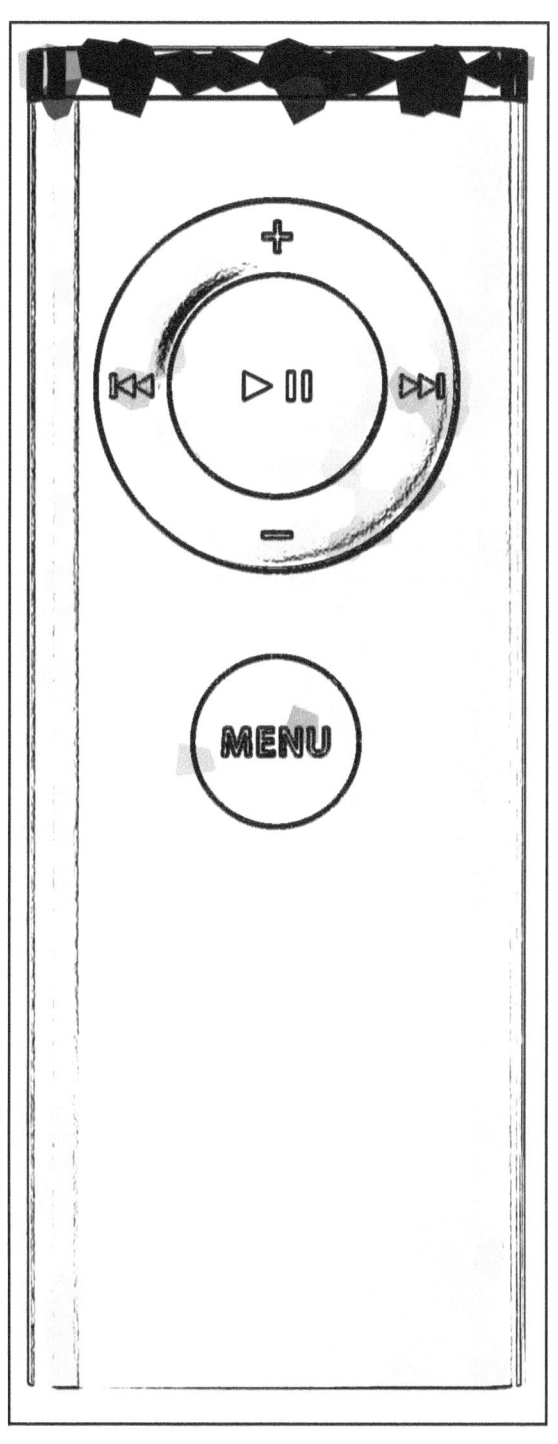

Architectural Minimalism intends to reduce complexities for the "externally visible distribution of responsibility" of the interactions, user interfaces, digital media, and artifacts. An architecturally minimal interaction, user interface, digital medium, and artifact can be created in a manner whereby simple elements are combined transparently; this thus provides a comprehensible and recognizable interface and interaction design that allows the user to understand and predict the interaction, user interface, digital medium, and artifact's behavior.

Compositional Minimalism intends to reduce complexities of the interface and interaction design through a "specificity for planned tasks" that extends the usefulness of the interactions, user interfaces, digital media, and artifacts beyond a single application or environment. A compositionally minimal interaction, user interface, digital medium, and artifact can be created in a manner that does not involve the user with unnecessary workflow.

Wireframes are visual guides that perform as screen blueprints for conceptualizing computer applications, digital media, and screen-based artifacts. Wireframes mainly focus on functionality and behavior of the content; hence they do not incorporate color, typography, or graphics. Depending on the levels of detail, wireframes can be divided into two categories: low-fidelity and high-fidelity. Low-fidelity wireframes include abstract, rough sketches, which incorporate less detail and are quick to produce. However, high-fidelity wireframes incorporate more details; hence they take longer to produce. The skeletal framework of a wireframe has three components: information design, navigation design, and interface design. The goal of wireframes is to provide a tangible feedback on the functionalities, identify navigation paths between pages, define placement and prioritization of information, and enhance ease of use and efficiency of the computer applications and digital media and artifacts.

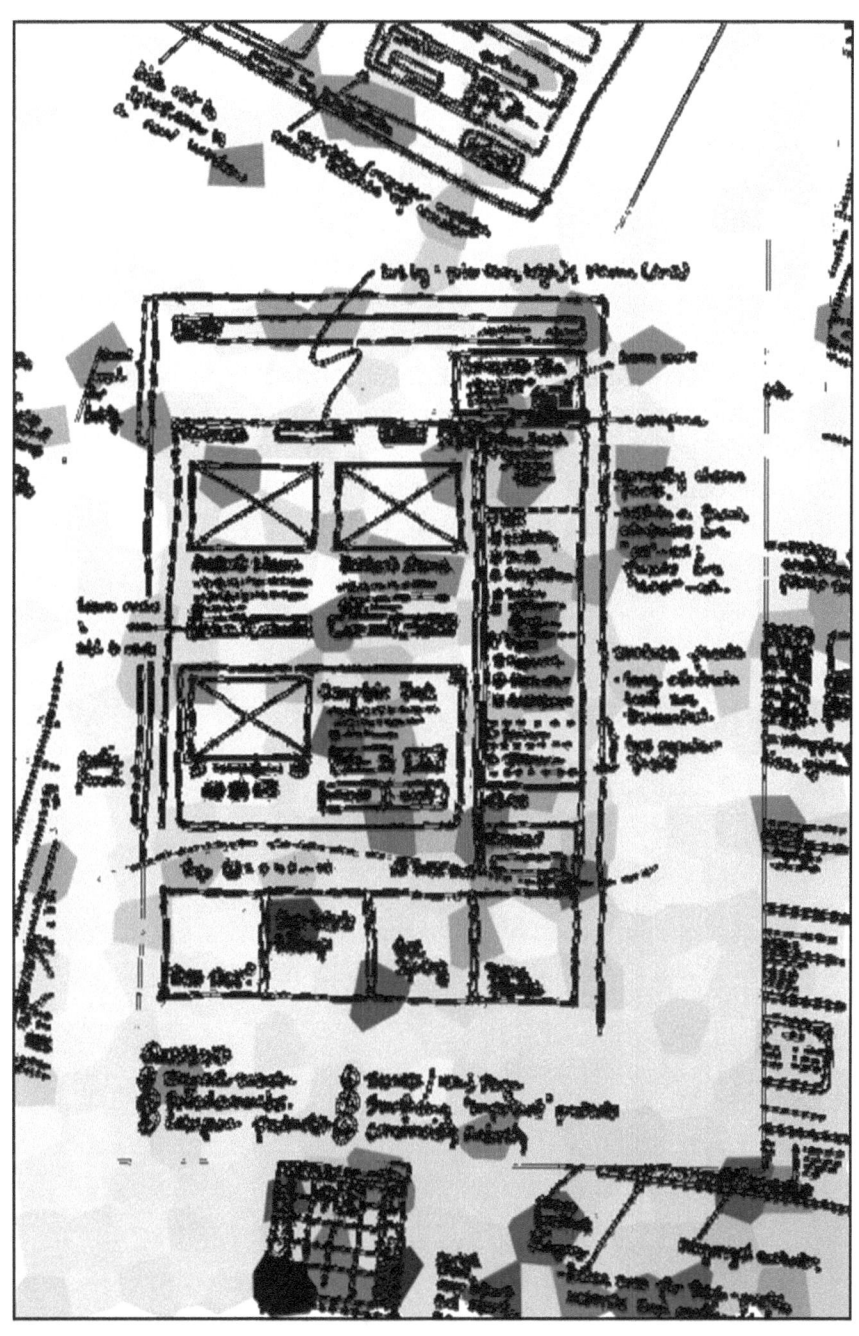

Storyboards consist of illustrations and images presented in sequence for previsualizing interactive media, motion graphic, motion picture, and animation. The process of planning and visual thinking of storyboards allows a group of people to participate in brainstorming more ideas together. Storyboards are mainly used in the fields of software, interactive media, film, comic books, novels, and business. The storyboarding process was developed at Walt Disney Productions during the early 1930s.

Persona is a fictitious character representing the typical user in a design scenario. The cast of characters represents the user population. To narrow down the spectrum of users, the primary persona should be chosen, who is the touchstone throughout the entire design process. Primary persona has a set of objectives that is different from those of the cast of characters. Primary persona is the personification of a naive and clueless user who is the primary focus of the design; hence, she/he has to be satisfied with the results of the end design. Making the primary persona happy would mean that each and every user would probably be happy. Thus, the end design should be created uniquely for the primary persona. Personas are introduced by Alan Cooper in his book *The Inmates Are Running the Asylum*, in 1999.

Jack	Sarah	Harry
Uses most phone features	Wants a simple phone, but functions as an integrated device	Will use almost all built-in mobile functionality
Uses phone to make, use contacts send texts and take pictures	Wants to easily read email and call back the sender	Will extend phone functionality with additional software
Always has mobile device with him	Needs "Popular" mail sever integration	Will look through and change change every menu option

Scenarios are the simplest form of prototypes that focus purely on narrating a single interaction session without any flexibility for the users. Scenarios can be used in early design and evaluation stages of products, services, and built environments in order to get users' feedback. Furthermore, scenarios can be used during the design of products, systems, and built environments to understand and envision users' interaction with them. Moreover, scenarios are task-oriented tools that are helpful for the participatory design process without any expenses of making a fully working system.

Use Cases are mainly diagrams that represent the detailed steps on how to achieve a goal via interaction between external actors and a product or system throughout the design process. The actor can be a person, technology, computer system, product, company, or organization. Use cases are normally represented in the form of a chart with two columns. First column shows that the actor, and second column shows the product or system. Furthermore, use cases are useful for identifying the level of work within a design process. Use cases can also show the users' needs and the goals that need to be accomplished within a product or system. However, use cases are often complex to write and understand. Moreover, use cases can be oversimplified or overemphasized, which may possibly lead to different interpretation or further imagination of action by different designers.

Prototypes and Mock-Ups are scale or full-size models of products, systems, and built environments. Prototypes and mock-ups are a faster and cheaper approach for iterative evaluation of products and systems with users. In prototypes and mock-ups the features and functionalities of products and systems are reduced to obtain feedback on the flow of interaction and ease of use while products and systems seem to be working, but they do not actually work. Vertical prototypes reduce the number of features. Horizontal prototypes reduce the number of functionalities. Furthermore, using a low-fidelity prototyping (such as paper prototypes) can quickly support the creation of simple mock-ups and at minimal cost. However, high-fidelity prototyping (such as HTML, CSS, and Adobe products) are more time-consuming and costly to produce. Prototypes and mock-ups are used in automotive devices, software engineering, system engineering, consumer products, and furniture.

Digital Paper Prototypes can be used to communicate the design concept, collect audience feedback on the flow of interaction, and work throughout a design. For the purpose of prototyping, different scenarios and personas and subsequent storyboards should be imagined and allowed to practice through combining the paper prototype and a digital medium. In general, the first and most important step in the prototyping process is to understand the audience and the intent of the prototypes' objectives. The second step begins with planning on a whiteboard or with paper and pencil and then drawing and drafting rapid iterative versions, reflecting incremental and evolutionary prototyping. The third step is to set expectations by determining the right level of fidelity and key functionality. The fourth step is to sketch out the idea and to write the labels. The fifth step is to make or fake the prototype by assembling a series of images and basic hide-and-show interactions built into the prototype. Finally, digital media can be used for creating basic interactivity by adding a few hotspots to each key frame.

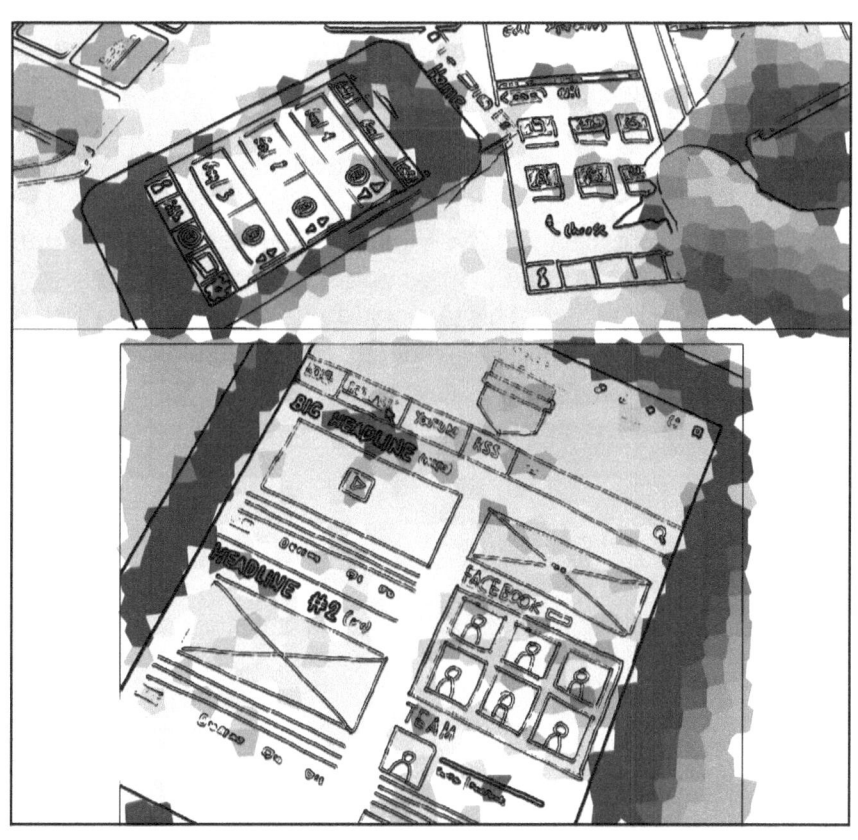

Cultural Probe is a technique that focuses on gathering subjective data about people's ideas, thoughts, values, and inspirations within the design processes. Cultural probes can include maps, postcards, cameras, or diaries. These artifacts, combined with evocative and inspirational activities, are given to participants to allow them to record specific events, feelings, and interactions. Thus, culture can be understood by eliciting people's inspirational responses. Cultural probe as a creative instead of analytical approach is often used in user experience research. This technique was developed by William Gaver, Tony Dunne, and Elena Pacenti in 1999, inspired by the art movement Situationist International.

Technology Probe is a method that can be used in the process of codesigning technologies with users. The goals of technology probes are to understand the needs, expectations, and limitations of the users, inspiring the users, designers, and researchers to envision new technologies, as well as evaluation of the technology with users. Technology probes focus on simple functionalities; however, they offer users flexible functionality choices. In addition, as opposed to prototypes, which are used iteratively later in the design process, technology probes are used in early phase of the design process.

Brainstorming is a participatory method for codesigning artifacts, tools, and technologies with a group of people. Brainstorming can be a useful method for exploring and discussing design questions raised throughout the design research. It is also an effective method for evaluating the design ideas and supporting the rationale in choosing between alternative concepts posed in design research. Brainstorming can also be useful in exploring if and how artifacts, tools, and technologies can enhance efficiency, simplicity, usability, and desirability.

Iterative Design is the process of redesigning and revaluating the products and systems to improve their usability challenges. Planning multiple iterations is beneficial for enhancing usability problems. In addition, considerable usability improvements will be achieved after early iterations. Furthermore, creating different iterations is a time-consuming and costly task. Hence, to prevent spending unnecessary time and money, major iterations that address the most important usability problems should be evaluated with the users.

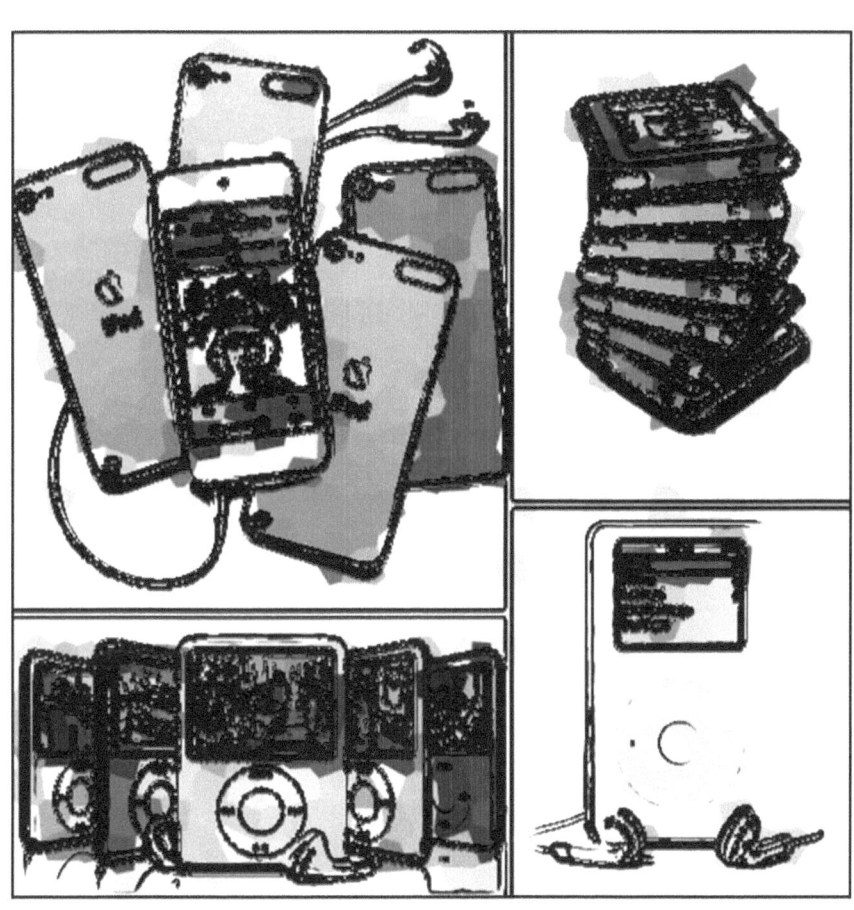

Conceptual Model is the most important element of a good design. Conceptual model represents the fundamental understanding held by a person about a subject or how something works. Users' conceptual model represents the models formed based on observation of physical existence, which occurs throughout the conceptualization process in mind. Users' conceptual model also represents human intentions and semantics. In addition, a good conceptual model allows users to predict the effects of their actions.

Design Specification defines the characteristics of products, systems, or built environments. Design specification also provides detailed information about the requirements of products, systems, or built environments. In addition, design specifications can be brainstormed and generated to address different dimensions of products, systems, or built environments. Design specification includes description and documents, such as drawings, sketches, dimensions, cost, maintenance, quality, safety, and environmental, ergonomic, and aesthetic factors.

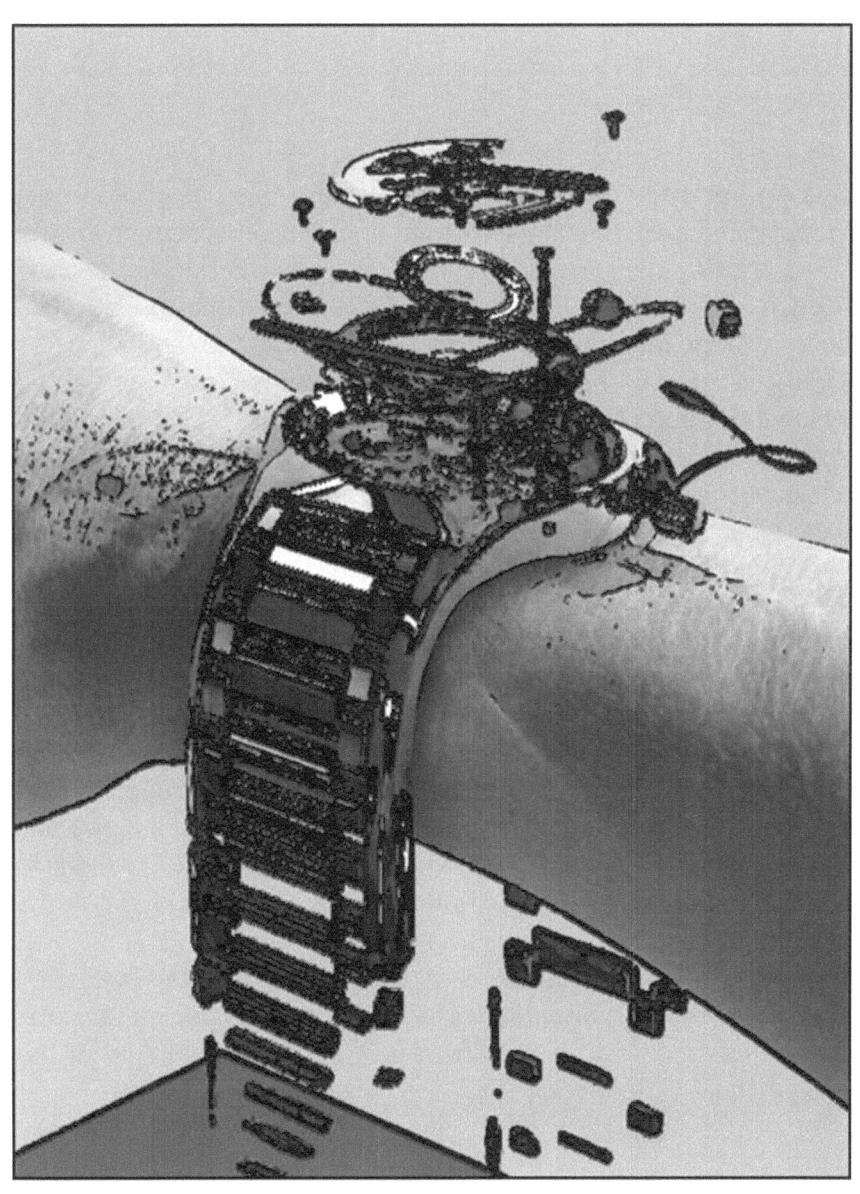

Design Variation is used to coordinate multiple design concepts to maximize commonality across a set of design concepts without compromising their individual characteristics. Design variation is used as a technique for creating different concepts based on a primary concept. Various concept characteristics are specified while changing forms and functions of a primary concept around which the group of design concepts are designed. This allows the achievement of a variety of concept characteristics, including form and function specifications and maximizing concept commonality. Hence, the primary design concept is common to all design variations. Commonality in concept is normally achieved through introducing a focal design goal that seeks to minimize the deviation of the input design specifications while satisfying the range of forms and functions. To compromise the tradeoff between satisfying the variation and maximizing concept commonality, each individual design concept is created around a common concept such that the design specifications for each design concept in the group are best satisfied. Furthermore, the fundamentals of design that define basic designs for each concept allow experiments with different options to achieve the optimal results in form and function integrated with the design concepts. Moreover, the design variables in concepts hold a small deviation constant to form the primary concept; thus, each design concept is compared against the previous design concept. Design variation can also show the process of evolution of the design concept from preliminary to final stages.

Do-It-Yourself (DIY) can be described as a design method that involves the use of personal skills for building, modifying, and repairing artifacts, technologies, houses, and fashions. DIY can also be described as a behavior where individuals construct, transform, and reconstruct artifacts, technologies, houses, and fashions by using raw and semiraw materials. In addition, DIY behavior can be motivated or inspired with various reasons, such as economic benefits, lack of product availability, lack of product quality, need for customization, craftsmanship, empowerment, community seeking, and uniqueness.

Form and Function are the fundamental building blocks in the design of objects and buildings. However, the debate on ornamental design and functional design has always been under the spotlight of modern industrial design and architecture in the twentieth century and onward. During this period of time, Louis Sullivan's phrase "Form follows function" was the adapted principle in modern products and buildings. Victor Papanek, who was an influential designer and design philosopher, was a proponent of "Form follows function." This principle was also adapted in the agile software development movement and aerodynamic automobile design. In addition, the Darwinian evolution argues that small variations in form would enhance the function in some parts of the population.

Conclusion

In the twentieth century, the modernist movement in design introduced the convergence of the design and science in search for scientific design products and processes. The design methods movement in 1960s attempted to scientize design on objectivity and rationality. During this period, Buckminster Fuller was associated with the design science revolution, and Herbert Simon developed the science of design. The design methodology continued to develop strongly throughout 1970s, 1980s, and 1990s. However, in 1980s, the differences between the epistemology of design and science became more evident. Glynn argued, "It is the epistemology of design that has inherited the task of developing the logic of creativity hypothesis innovation or invention that has proved so elusive to the philosopher of science." Furthermore, there remain some interpretations of relationship between the science and design. They include scientific design, design science, science of design, and design as a discipline. Scientific design refers to the modern practice of design, which is based on scientific knowledge, and it is improved from preindustrial and craft-oriented design. Design science attempts to formulate a systematic and rationalized method for design, which approaches the design as a scientific activity. Science of design attempts to provide an understanding of design through scientific inquiry. Design as a discipline attempts to develop its own independent intellectual approaches to design theory and research. Moreover, similar to sciences and arts, which focus on scientific and artistic forms of knowledge, design as a discipline should concentrate on forming the independent knowledge drawing upon the stronger histories of inquiry in the sciences or the arts.

References

As the term *design knowledge* suggests notions of creative inquiry, it relates to a wide range of ideas covered by design research process. For the purpose of further reading, the selected texts listed here cover the relevant intellectual design issues, often making particular reference to sciences or arts. These texts provide a route through the multifaceted subject of design knowledge while also being fascinating for their views and interpretations of sciences or arts.

J. Baldwin, *Bucky Works: Buckminster Fuller's Ideas for Today*, New York: Wiley, 1996.

H. Beyer and K. Holtzblatt, *Contextual Design: Defining Customer-Centered Systems*, San Francisco: Morgan Kaufmann Publishers, 1998.

M. E. Bratman, *Facts of Intention*, Cambridge: Cambridge University Press, 1999.

A. Crabtree, *Designing Collaborative Systems: A Practical Guide to Ethnography*, London: Springer, 2003.

J. W. Creswell, *Qualitative Inquiry and Research Design: Choosing among Five Approaches*, California: SAGE Publication, 2007.

N. Cross, *Design Thinking: Understanding How Designers Think and Work*, Oxford: Berg/Bloomsbury Academic, 2011.

N. Cross, *Engineering Design Methods: Strategies for Product Design*, Oxford: Wiley, 2008.

N. Cross, *Designerly Ways of Knowing*, Boston: Birkhauser, 2006.

N. Cross, "Designerly Ways of Knowing: Design Discipline versus Design Science," *Design Issues,* vol. 17, no. 3, pp. 49–55, 2006.

N. Cross, "Can a Machine Design?" *Design Issues,* vol. 17, no. 41, pp. 44–50, 2001.

N. Cross, "Natural Intelligence in Design," *Design Studies,* vol. 20, pp. 25–39, 1999.

N. Cross, H. Christiaans, and K. Dorst, *Analysing Design Activity*, Oxford: Wiley, 1996.

A. Cooper, *The Inmates Are Running the Asylum: Why High-Tech Products Drive Us Crazy and How to Restore the Sanity*, Indianapolis: Sams Publishing, 2004.

H. Dreyfuss, *Designing for People*, New York: Allworth Press, 2003.

C. Emden and M. Sandelowski, "The Good, the Bad, and the Relative, Part One: Conceptions of Goodness in Qualitative Research," *International Journal of Nursing Practice,* vol. 4, pp. 206–212, May 1998.

C. Emden and M. Sandelowski, "The Good, the Bad, and the Relative, Part Two: Goodness and the Criterion Problem in Qualitative Research," *International Journal of Nursing Practice,* vol. 5, pp. 2–7, 1999.

T. Flew, *New Media: An Introduction*, South Melbourne, Victoria: Oxford University Press, 2005.

W. W. Gaver, "Situating Action II: Affordances for Interaction: The Social Is Material for Design," *Ecological Psychology,* vol. 8, no. 2, pp. 111–129, 1996.

S. Glynn, "Science and Perception as Design," *Design Studies,* vol. 6, no. 3, pp. 122–126, 1985.

J. Hollan, E. Hutchins, and D. Kirsh, "Distributed Cognition: Toward a New Foundation for Human-Computer Interaction Research," in *ACM Transactions on Computer-Human Interaction,* 2000.

K. Holtzblatt, J. B. Wendell, and S. Wood, *Rapid Contextual Design: A How-To Guide to Key Techniques for User-Centered Design*, San Francisco: Morgan Kaufmann Publishers, 2005.

E. Hutchins, "Distributed Cognition," in *The International Encyclopedia of the Social and Behavioral Sciences,* California, 2001.

E. Hutchins, *Cognition in the Wild*, Cambridge, Massachusetts: The MIT Press, 1995.

H. Hutchinson, W. Mackay, B. Westerlund, B. B. Bederson, A. Druin, C. Plaisant, M. Beaudouin-Lafon, S. Conversy, H. Evans, H. Hansen, N. Roussel, B. Eiderbäck, S. Lindquist, and Y. Sundblad, "Technology Probes: Inspiring Design for and with Families," in *CHI*, Florida, 2003.

P. M. Kroonenberg, "Three-Mode Principal Components Analysis of Semantic Differential Data: The Case of a Triple Personality," *Applied Psychological Measurement,* vol. 9, no. 1, pp. 83–94, March 1985.

R. Larson and M. Csikszentmihalyi, "The Experience Sampling Method," *New Directions for Methodology of Social and Behavioral Science,* vol. 15, pp. 41–56, 1983.

Y. S. Lincoln and E. G. Guba, *Naturalistic Inquiry*, California: Sage Publications, 1985.

J. H. Murray, *Inventing the Medium: Principles of Interaction Design as a Cultural Practice*, Cambridge, MA: MIT Press, 2012.

J. H. Murray, *Hamlet on the Holodeck: The Future of Narrative in Cyberspace*, Cambridge, Massachusetts: MIT Press, 1997.

D. A. Norman, *The Design of Everyday Things*, New York: Basic Books, 2013.

D. A. Norman, *Living with Complexity*, Cambridge, Massachusetts: The MIT Press, 2010.

D. A. Norman, *Things That Make Us Smart*, Massachusetts: Addison-Wesley Publishing Company, 1993.

H. Obendorf, *Minimalism: Designing Simplicity*, London: Springer, 2009.

K. O'Doherty and E. Einsiedel, *Public Engagement and Emerging Technologies*, Vancouver, BC: University of British Columbia Press, 2012.

V. J. Papanek, *Design for the Real World: Human Ecology and Social Change*, New York: Pantheon Books, 1971.

Y. Rogers, H. Sharp, and J. Preece, *Interaction Design: Beyond Human-Computer Interaction*, Sussex: Wiley, 2011.

J. Rudd, K. Stern, and S. Isensee, "Low- vs. High-fidelity Prototyping Debate," *Interactions,* vol. 3, no. 1, pp. 76–85, January 1996.

M. Sandelowski, "Focus on Qualitative Methods: Sample Size in Qualitative Research," *Research in Nursing & Health,* vol. 18, pp. 179–183, 1995.

R. Sefelin, M. Tscheligi, and V. Giller, "Paper Prototyping—What Is It Good For? A Comparison of Paper- and Computer-Based Low-Fidelity Prototyping," in *CHI 2003*, Florida, 2003.

D. Schön, *The Reflective Practitioner: How Professionals Think in Action*, London: Temple-Smith, 1983.

H. A. Simon, *The Science of the Artificial*, Cambridge: MIT Press, 1996.

J. Short, E. Williams, and B. Christie, *The Social Psychology of Telecommunications*, London: John Wiley & Sons Ltd., 1976.

D. Schuler and K. Namioka, *Participatory Design: Principles and Practices*, Hillsdale, NJ: Lawrence Erlbaum Associates, 1993.

L. A. Suchman, *Human-Machine Reconfigurations: Plans and Situated Actions*, Cambridge: Cambridge University Press, 2007.

L. A. Suchman, *Plans and Situated Actions: The Problem of Human-Machine Communication*, Palo Alto, California: Xerox Corporation, 1985.

T. Tufte and P. Mefalopulos, *Participatory Communication: A Practical Guide*, Herndon, VA: World Bank Publications, 2009.

T. Z. Warfel, *Prototyping: A Practitioner's Guide*, Brooklyn, NY: Rosenfeld Media, 2009.

Index

I

inductive, bottom-up reasoning, 57
industrial design, 33, 81, 123
Inmates Are Running the Asylum, The, 97, 128
instrumental case study, 15
interdisciplinary design studies, 1
interior design, 33
Interpretation Methods, 57
interview, 35, 53, 55
intrinsic case study, 15
Iterative Design, 113

J

journaling, 55

K

K-J methods, 57

L

landscape architecture, 33
Larson, Reed, 27, 129
Likert scale, 39
low-impact materials, 33

M

media design, 81
memoing, 55
methodology, 7, 27, 37, 125, 129
minimalism, 83, 85, 87, 89, 91, 130

model consolidation, 19
modern industrial design and architecture, 123
modernist movement in design, 125

N

naturalistic generalizations, 81
natural resources, 33

O

objectivist, 3
observation, 29, 37, 41, 59, 115

P

Pacenti, Elena, 107
Papanek, Victor, 123, 130
participatory design, 21, 99, 131
perceived access structure, 87
persona, 25, 97
Physical Model, 79
pilot study, 35, 45
postmodernism, 5
preindustrial, 125
probable truth, 5
prototypes and mock-ups, 19, 105
psychology, 17, 128, 131
psychometric scaling, 39

Q

qualitative research, 3, 5, 15, 53, 55, 81, 128, 130
quantitative research, 3

R

Rapid Contextual Design (RCD), 19, 129
reductionism, 83
Reflective Practitioner, The: How Professionals Think in Action, 1, 130
renewability, 33
Research-in-the-Wild, 13
rhetorical situation, 25
robust eco-design, 33

S

Scandinavian, 21
scenarios, 99, 105
Schön, Donald, 1, 130
science of design, 125
scientific design, 125
scientize design, 125
semantic differential scale, 39
semantics, 115
sequence model, 73
seven quality processes, 57
Simon, Herbert, 125, 131
Situationist International, 107
socially oriented, 11
socioeconomic, 33
sociology, 17
storyboards, 95, 105
Structural Minimalism, 87
subjectivist, 3
Sullivan, Louis, 123
survey, 27, 39
sustainable design, 33

T

task analysis, 67
Theoretical Perspective, 5–6
Thinking Aloud, 47
triangulation, 51

U

universal design, 31
urban design, 33, 81
urban planning, 33
use cases, 101
User-Centered Design (UCD), 25, 29, 37, 83, 129
user empowerment, 21
user environment design (UED), 19
user experience design (UXD), 23

V

variations in design, 119, 123
vertical prototype, 103

W

wireframes, 93
work-in-context, 13
written materials and physical artifacts, 55

www.ingramcontent.com/pod-product-compliance
Lightning Source LLC
Chambersburg PA
CBHW030756180526
45163CB00003B/1046